4 Short Topics in
System Administration

Educating and Training System Administrators: A Survey

By David Kuncicky and Bruce Alan Wynn

Published by the USENIX Association for
SAGE, the System Administrators Guild
1998

ISBN 1-880446-99-5

Copies of these publications are available to members of SAGE for $5.00 and to non-members for $7.50. Outside the US and Canada, please add $3.50 per copy for postage (via printed matter).

For copies and for membership information, please contact:

The USENIX Association
2560 Ninth Street, Suite 215
Berkeley, CA 94710 USA
Email: *office@usenix.org*
Web: *http://www.usenix.org*

First Printing, 1998

Printed in the United States of America, on 50% recycled paper, 10-15% post-consumer waste.

Acknowledgments

This document has benefited from review by a number of people who generously donated their time and expertise. The authors thank the following for their detailed and helpful suggestions: Eric Berglund, David Jones, Evi Nemeth, and David Parter.

Preface

As any practice makes the transition to a profession, nothing could be more important that a proper education. System Administration is one of the newest professions in the world of advanced technology, and has rapidly become one of the most critical. One of the goals of the Short Topics in System Administration series is to enhance recognition of system administration as a profession. Without such recognition on a large scale, there is little motivation for the establishment of formal educational programs.

I hope that this booklet will serve as a catalyst for more focused discussion and communication concerning the proper education of our future system administrators. The authors have spent years studying this issue from two different perspectives: collegiate and commercial. The results of their study provide insight in to the current state of education and should be the seed for further ideas. Education is key to meeting the demands that will be placed on this critically important industry.

William LeFebvre
Alpharetta, Georgia

Table of Contents

1.0 Introduction

The purpose of this booklet is to describe the state of training and education for system administrators and to explore ways that the process can be improved. The discussion of this topic is central to the mission of SAGE in its effort to establish standards of professional excellence and propagate knowledge of good practice in the profession. It is an important topic for employers of system administrators, for students of system administration, and for educators who are trying to develop a curriculum in this area. This booklet is both a report of the current state of affairs and a call to action for educators and system administration professionals.

The discussion of the training and education of system administrators begins with a description of the skills and knowledge required for the job of system administration. The first publication in the SAGE Short Topics Series, *Job Descriptions for System Administrators* will be used as a starting point for this discussion [1]. The *Job Descriptions* booklet describes the required skills and background for the system administration profession.

System administrators can acquire the knowledge to do their jobs through formal education, commercial training, and independent study. This classification method is not perfect but will serve the purpose of guiding the discussion. Within each method, the current state of affairs will be described and examples of existing programs will be given.

Two case studies will be presented: one from academia and one from the commercial sector. The analysis of these two cases provides examples of the difficulties, successes, and failures that can occur in the development and initial delivery of system administration programs.

1.1 The Current State of Affairs

The means of acquiring system administration knowledge and skills are widely varied. Few system administrators are trained as such through formal academic programs. Many are self-taught. Others have learned skills through a combination of vendor-based training and apprenticeship.

What is wrong with the status quo? After all, thousands of networked computing sites are in operation. System administrators are being trained by some means.

Several general trends are increasing pressure on current knowledge delivery methods. One trend is that the demand for competent system administrators exceeds the supply. As system administrators, we may like a seller's market; from a corporate viewpoint, the trend is disastrous. A corporation will have a difficult time marketing its computer or networking systems if there is no pool of talent available to operate and maintain them. The nation's chronic shortage of information technology professionals is escalating, especially in system administration-related areas such as client-server technologies, internetwork administration, UNIX, and NT administration.

A second trend affecting the commercial sector's ability to train its employees is the increasing internationalization of corporations. Employees are spread out over the globe and it is becoming less feasible to transport and train them at central facilities.

The following sections explore the problem from the viewpoints of the major players: the student, the educator, and the employer.

1.2 The Student's Point of View

The student who wishes to become a system administrator today will have a difficult time locating advice about comprehensive education and training in the area. Students have no way of knowing what degrees, if any, to pursue, what foundation courses to take, or how to find an adequate apprenticeship. This is partly due to a lack of such programs, especially in post-secondary institutions, but it is also due to a lack of consolidated information about existing programs.

Many vendor-based short courses are available . However, it is difficult for students to determine if courses from different vendors overlap or if they complement each other.

There is a growing trend towards product-specific certification of system and network administrators such as the Cisco CCIE, Microsoft Certified Professional, and Novell CNE certifications. However, these are less than satisfactory for the student who wishes to become a system administration professional. There appears to be little coordination of curriculum among vendors. Vendor-based courses tend to be narrowly focused and may not address the issues of integrating multiple technologies into a single installation. A vendor-based course may present solutions that are biased towards a particular vendor's strengths.

1.3 The Educator's Point of View

The educator who is trying to build a program in system administration will find little guidance. This is especially true in academia, where system administration is just beginning to be accepted as a legitimate focus area. Vendor-based training programs have a longer history, and more prototypes are available.

It should be noted that educators from the commercial sector and educators from academia have different agendas. Historically, a tension has existed between corporate training and higher education. The separation has been by culture and design. Corporations want to control training content. They need the training to be immediately applicable to the job at hand. They want to deliver the training cheaply, deliver it in short bursts, and be finished with it so that employees can move on to productive endeavors. Higher education programs, on the other hand, operate in broader terms. The content is principle-driven and is not required to have immediate applicability. The educators themselves are not driven to shorten the educational process.

A vendor-based program that does not incorporate or build upon general computing and networking fundamentals will produce a system administrator who tends to be myopic and rigid. Academic programs that teach only theory and generalities will produce graduates who "talk a good game" but can't perform.

In an ideal situation, a program would provide a solid background through a degree in computer science or a related field. The program would include an intensive practical internship in a heterogeneous computing environment, either during or immediately after the course of study. The graduate would then be immediately employable in the areas in which the practicum occurred and would be able to learn new systems quickly using the fundamental knowledge acquired in the degree program. The educator has the difficult task of balancing the general and specific when designing a program in system administration.

1.4 The Employer's Point of View

As is often the case in emerging professions, today's first-line managers generally don't fully understand the field of system administration, and they don't really know what skills are important or useful. Consequently, system administrators are often hired based on best estimators - college degree and years of experience.

The years of experience shown on a system administrator's resume can be misleading. The years of experience that a person has accumulated are meaningful only if the experience is relevant. Because the field is so broad and so rapidly changing, ten years of experience in one environment may be nearly useless in another environment. The manager attempting to hire a system administrator has no mechanism to assess the applicant's relevant skill set. Some personnel systems, especially in government, require that the manager use the applicant's degree information and years of experience as selection criteria. In a worst case, the manager may be forced to hire a person to design the company's firewall who has spent ten years changing tapes on the graveyard shift.

Resumes are less than perfect sources of information. Often employers will look for the appearance of well-known acronyms (TSO, ATM, NFS, DNS, SQL, EDI, VHDL, for example) and will assume that their presence implies well-defined skill levels. Applicants learn to toss around acronyms to employers, and interviews ensue that consist primarily of compu-babble. One amusing method that can be used to weed out such behavior is to ask applicants to describe their experience with a phony acronym. Evi Nemeth has such a question on her system administration examination on the CD accompanying the *Handbook of System Administration* [2].

2.0 The System Administration Job

2.1 Introduction

Today's system administrator must possess a broad technical skill set. The acquisition and assimilation of these skills require a strong and current background in areas such as operating systems, networking protocols, and hardware platforms. On top of this fundamental body of knowledge, the system administrator learns and practices a variety of technical skills.

A system administrator may be required to perform technical tasks as varied as polishing a fiber optic cable, diagnosing network bottlenecks, installing system software, and planning the configuration of a large server group. General problem-solving skills and the ability to multitask under pressure are required.

Additionally, the system administration role requires good interpersonal skills, written and verbal communication skills, task and time management, and political shrewdness. A system administrator may be required to perform nontechnical tasks as varied as teaching classes and writing strategic plans. Increasingly, the system administrator must be aware of legal issues such as right to privacy, encryption laws, and harassment.

One method of looking at the system administration job is to outline the skill set needed at different levels of expertise. The best source in this area is the Short Topics in System Administration, Volume One: *Job Descriptions for System Administrators,* edited by Tina Darmohray and published by SAGE [1]. This booklet lists templates for core job descriptions for system administrators. It serves as an excellent task-oriented coverage of required skills for the profession.

In the booklet, system administration job descriptions are divided into four levels of expertise:

- Novice
- Junior
- Intermediate/advanced
- Senior

The minimum technical and nontechnical skills expected at each level are described. In addition, the required and desired backgrounds for each level are listed.

2.2 Technical Skill Requirements

Job Descriptions focuses on the UNIX operating system, but the technical skill categories may be generalized to other systems. A well-rounded system administrator needs these technical skills:

- Proficiency with basic system administration tools
- Proficiency with the installation and administration of operating systems, mail systems, networked printing, and other appropriate application software
- An understanding of the relevant operating system's methods of managing processes, interprocess communication, memory, devices, and file systems
- An understanding of distributed networked computing concepts, and proficiency with routing, network management, network file services, network naming services, and client-server configuration
- Proficiency with authentication schemes, network security, backup methods, and physical security
- Ability to program in an administrative scripting language
- Ability to modify and debug C programs
- Proficiency in the installation, diagnosis, and repair of computing devices down to the board level
- Proficiency in installation and resolution of hardware problems with network cabling and network devices such as transceivers, routers, and switches

Job Descriptions describes these technical tasks in detail and provides examples of each.

2.3 Nontechnical Skill Requirements

Many nontechnical skills are also required to be an effective system administrator. These include:

- Good interpersonal relations
- The ability to communicate effectively in written and verbal forms
- The ability to multitask under pressure
- Problem-solving skills
- A dogged determination to find solutions to problems
- The ability to test solutions
- The discipline to complete tasks

A system administrator may also be expected to perform tasks requiring managerial skills such as:

- Supervising and managing technical staff
- Writing system design and upgrade proposals
- Writing bid specifications, managing bid processes and contracts
- Training users
- Managing a help desk
- Being on call
- Writing system documentation
- Writing and administering system policies

2.4 The Educator's Task

The stated education and background for system administration jobs are usually:

- A degree in computer science or related field
- Experience

Most of these listed technical and nontechnical skills are not taught in the typical computer science program. The majority of the skills are learned after graduation. Training for these skills may consist of on-the-job training, an informal apprenticeship, or a series of vendor-based short courses. It would be advantageous if the bulk of the skills necessary for the job of system administration were taught during the student's course of study. Instead, the current situation requires the student to learn most of the required job skills in an ad hoc fashion after obtaining a degree!

The task for educators is to organize these skill sets into course content and locate or create courses that teach the skills, convey fundamental approaches, and provide experience. This does not mean that educators should be tempted to design "cookbook" courses in which students are taught the steps of a process without being taught the underlying reason for those steps. In the field of system administration, such an approach tends to fail.

There are several reasons for this. One, there are too many specific recipes to learn - too many vendors, too many versions, and too many environments. The new system administrator must be prepared to learn new procedures and solve new problems constantly. A general knowledge of principles can be applied to many tasks, but simple rote learning of specific tasks will not necessarily generalize.

For example, if one were looking for a Perl programmer, it is much better to hire an employee who has a fundamental understanding of programming languages than an employee who knows Perl syntax but nothing else. The former will quickly learn Perl over a weekend. The latter will stumble when it is necessary to change from Perl to another scripting language.

A difficult problem for educators is that diagnosis and problem solving are procedural processes. Methodically diagnosing and solving problems is a key to good system administration. Students who memorize solutions rather than create solutions end up being support junkies. That is, they tend to try several random solutions and then, failing to solve the problem, they call the software support line for help. It is preferable to teach students problem-solving methods and then give them the opportunity to confront difficult diagnostic problems in real-world conditions. See Dave Jones's System Emergency Week for an example [3].

Some of the nontechnical skills such as good interpersonal relations, the ability to multitask under pressure, and problem solving are difficult to teach, but a program in system administration can help foster some of these skills. Specific courses in interpersonal communication can be included as part of a curriculum. Tutors and instructors can act as role models during internships or during the completion of competency apprenticeships. A program can act as a filtering mechanism, gently guiding people who are not suited for a service profession away from the field.

3.0 Formal Education Programs

3.1 Introduction

The profession of system administration is younger than many other professions that require similar levels of skill and responsibility and that offer similar monetary rewards. One of the attributes of a young profession is that its members tend to acquire jobs based solely on the knowledge and skills that they can demonstrate - with little emphasis on the degrees, certificates, and/or licenses they hold. "Because system administration draws on knowledge from many fields, and because it has only recently begun to be taught at a few institutions of higher learning, system administrators may come from a wide range of academic backgrounds. Most get their skills through on-the-job training by apprenticing themselves to a more experienced mentor." [1] Many system administrators were educated and trained in other professions. A common story is that of a professional who is employed in a nonsystem administration position and is assigned the task of administering the department's computing system.

The profession of system administration has reached a stage where industry and government demands much from system administrators: technical competence, trust, high ethical standards, communication skills, and personnel management skills. At the same time, the gap between supply of system administrators and demand for system administrators is steadily growing. It is time for the profession to consider and promote the formal education of system administrators to help insure that standards of competence in the field are maintained as the supply of system administrators increases.

As a profession matures, it tends to become formalized in a number of ways:

- The methods of transferring the profession's knowledge and skills become codified (curriculum design, accreditation, etc.)
- Professional organizations develop
- Some method of certification and/or licensure develops
- Responsibility for liability increases
- The profession becomes subject to regulation

Some professions welcome the formalization process because it gives the profession legitimacy (e.g., massage therapy, chiropractic). For some professions, formalization opens financial doors such as third-party insurance payments (e.g., social work, nursing, counseling). One would suspect that some system administrators chose their field partly because it does *not* have many of these characteristics. System administrators are a spirited and freewheeling bunch. Good system administrators possess a broad array of valuable skills that gives them job security and allows them to avoid licensure, dress codes, and nine-to-five work hours.

The authors are neither advocating nor rejecting rapid or forced changes to the profession of system administration. We are merely noting a process that tends to occur as a profession matures.

This section explores some of the attempts to formalize the education of system administrators in post-secondary educational institutions. The education of system administrators is an area that we believe *does* need to be developed and structured. There is a large enough body of general knowledge in the area to support a specialization of system administration in the curricula of computer science, computer engineering, and information science departments.

Some colleges and universities offer system administration courses. A list of some of these courses is presented in Appendix A. Most of these courses are offered within the curricula of departments of computer science, computer engineering, and management information systems. Some are offered as certificate courses; that is they are not offered for credit towards a degree but a certificate is conferred at the end of the course.

Most of the credit courses are offered as electives. Few programs incorporate system administration courses into their core curricula. Some programs that offer a more comprehensive approach are:

- The Department of Computer Science at the University of Colorado at Boulder
- The Bachelor of Information Technology degree at the Central Queensland University
- The Computer and Network System Administration Master's Track at Florida State University

3.2 Curriculum Design

3.2.1 Terminal Degree Level There has been some resistance to the addition of system administration into the computer science curriculum. Instructors of system administration courses reflect this resistance with comments such as, "Most large universities are more intent on teaching the theory behind how something is done, rather than providing an apprenticeship on how to do it." There is some concern in academia that universities are becoming high-level vocational schools. Notable examples of fields that confer professional degrees and still manage to build a body of scientific knowledge are psychology, medicine, business, and social work. If colleges and universities do not produce computer scientists and engineers with the skills that society needs, then the programs will be seen as increasingly irrelevant and the needs will be satisfied elsewhere.

Some may argue that the body of knowledge and skills called system administration is too diverse, too rapidly changing, and too experiential in nature to be formalized in a college-level curriculum. The same arguments could be made for the field of medicine. The physician must undergo continuous retraining and education in an era when new research rapidly changes diagnostic procedures and technological progress frequently requires learning new treatment methods. The graduate of a program in system administration should also expect to pursue a lifetime of continuing education and training to maintain a current level of knowledge and skills.

What is the appropriate level for a curriculum in system administration: certificate, vocational school, community college, four-year college program, or graduate program? The answer is that system administration should be taught at all of these levels. A comparative example is the area of programming languages. The subject of programming languages is taught at many levels including:

- Short courses (noncredit)
- Secondary schools
- Vocational schools
- Community colleges
- Undergraduate four-year programs
- Graduate programs

The scope of knowledge and the expectations for programmers who successfully complete these programs varies widely. The graduate of a vocational program may learn a single language. The graduate of a four-year program learns several languages and studies the underlying concepts of programming language design and implementation such as control structures, data structures, and type checking. The student in a graduate program is taught the formal analysis of programming language syntax and semantics e.g., axiomatic semantics and denotational semantics.

A similar separation of educational levels is appropriate for system administrators. The student who needs to know only how to maintain a single type of PC-based network or who will administer a single vendor's mainframe might appropriately complete a vendor's short course or attend a vocational training program.

The student who will maintain a heterogeneous network of clients and servers will be better prepared after graduating from a two-year or four-year program. This student will have a broader background in operating systems, computer architecture, computer programming, and networking. The more educated student will have a greater ability to survive technological change and is likely to be less resistant to paradigm shifts in operating systems, networking, and computer architecture. The vocational school graduate will need to be retrained more often and may be more resistant to technological change. An example is the mainframe system administrator who resists the move to client-server computing.

The student who pursues a graduate degree will have all of the skills and knowledge of the four-year graduate, but, in addition, will be prepared to teach, perform research, and write scholarly articles. Nontenure faculty positions have been offered at colleges and universities to system administrators who hold a minimum of a master's degree.

3.2.2 Undergraduate Curriculum An undergraduate track in system administration would include several new courses in addition to the standard computer science or computer engineering undergraduate requirements. It is assumed that the student first obtains a strong background in operating systems, computer architecture, computer networks, and several programming languages, including C and C++. Several types of courses currently offered are candidates for new core courses in an undergraduate curriculum in system administration. These include:

- Computer system administration
- Computer networks with applications
- A system administration seminar

Examples of each of these course categories are provided in the following sections.

In addition to course work, the program should offer an extensive experiential component including an internship or other job-related activity. This should be supplemented by extensive lab experience or the completion of a competency checklist. The addition of an experiential component to the curriculum is detailed in the following sections.

3.2.3 Graduate Curriculum A graduate curriculum would include graduate courses in:

- Operating systems
- Data and computer communications
- Introductory seminar on research
- Computer networks with applications
- Computer architecture
- Computer and network system administration
- A system administration seminar
- The successful defense of a suitable thesis or master's project

The graduate practicum could be an extended version of the undergraduate practicum. The next two sections describe examples of new core courses in a system administration curriculum.

3.3 The System Administration Course

The authors surveyed system administration courses currently taught as electives at colleges and universities. Some of the recent courses with Web access are listed in Appendix A. One nice example is Dave Jones's course at Central Queensland University [3]. The course was first taught as an elective in 1992 and integrated into the undergraduate curriculum in 1994. Several papers describe the evolution of the course and program at CQU [4][5]. The survey of system administration courses reveals some generalities that may be helpful to others who are developing a similar course.

3.3.1 Prerequisites The most common prerequisites for the surveyed courses include:

- Operating systems
- User-level knowledge of UNIX
- C programming

There are at least two ways to view the placement of this course in the curriculum. One view is that the course should be offered as early as possible so that the student has an opportunity to apply the learned knowledge for as long as possible before graduation. The other view is that the course should be offered at the senior level. This is based on the notion that the practical knowledge and skills taught in the course will be assimilated better after fundamental principles have been learned. In the second case, other prerequisites might be courses in computer networking and computer architecture. Examples of system administration topics that are affected by these additional prerequisites are performance analysis, network monitoring, and security.

3.3.2 Textbooks The most popular textbook for UNIX system administration courses is *The UNIX System Administration Handbook* by Evi Nemeth et al. [2]. Other textbooks used in the surveyed courses are listed in the Reference section [6][7][8]. Dave Jones has placed his entire course notes on the Web and does not require a paper textbook [5].

3.3.3 Course Content A short description of the content for many of the system administration courses reads roughly like the table of contents for Evi Nemeth's book:

- Daemons and services
- Booting the system

- Process management
- Adding new users
- Local disk and filesystem management
- Network file system management
- Configuring a kernel
- Devices and drivers
- Serial devices
- TCP/IP and routing
- Network hardware
- DNS configuration
- SLIP and PPP
- Disk space management (quotas)
- Sendmail
- Performance analysis
- System accounting
- System logging (syslogd)
- System security
- Intrusion prevention and detection
- Policies and procedures

Some instructors are adding Windows NT components to the course content. Some, such as Jeff Bauer's course at FSU, are adding UNIX/NT interoperability [9]. The students are required to install and configure Linux on one machine and Windows NT server on another. The students then install and manage services that span both operating systems.

3.3.4 Labs and Equipment For a system administration course to be effective, the students must have hands-on experience. This includes having administrator privilege and access to the system that is not granted to other students in a general-purpose computer lab. This implies that the student is given the ability to trash the operating system. Linux (a free UNIX-like operating system) has been a salvation for instructors. Before Linux it was difficult for instructors to allocate enough workstations to allow each student root access. Some courses require students to install Linux on the student's own PC. It would be nice to build a lab specifically for the system administration course, but this may not be possible.

One effective method is to set up PCs to dual-boot. The default boot is MS Windows or Linux, which is installed for general lab use. The secondary boot partition is allocated to an individual student who uses it to explore the system with root privileges. When finished with a system administration session, the student reboots the machine to the default OS and it is left ready for others to use. This method has security implications that must be carefully considered.

3.3.5 Assessment The assessment of system administration students ideally will include the evaluation of problem-solving skills and a performance test in addition to the usual written exams. The System Emergency Week model introduced by Dave Jones has worked well at FSU. In this method, the instructor intentionally creates a system emergency by "breaking" the system in some way. The student is expected to diagnose the problem and present a solution within a fixed time frame. If, after an initial period, the problem has not been solved, the student may confer with other students and elicit their help. The student is asked to think aloud during the entire process. The pressure of having the instructor and several others look over the student's shoulders provides real-world tension.

3.4 The Networks with Applications Course

Many college-level network courses consist of a survey of the OSI model and network protocols and standards such as the TCP/IP suite, Ethernet, Token Ring, and ATM. The network course for system administrators should include additional components. It may be that a two-course sequence is needed provide this additional coverage.

Additional components are:

- Exposure to network hardware and network software tools
- Projects that require setting up and administering local area networks,
- A programming component that involves building network tools and client-server applications

3.4.1 Example #1: Network Systems Course An example of a course that meets these new requirements is the network systems course taught by Evi Nemeth at the University of Colorado [10]. The course was co-listed as a senior undergraduate and beginning graduate course. The course offered hands-on experience with network equipment such as a network analyzer and included timely, real-world projects such as configuring a Cisco router. Software projects using BSD sockets and Sun RPC were required. Graduate students were assigned additional research requirements.

The prerequisites were:

- Undergraduate or graduate operating systems
- Permission of the instructor

- A thorough knowledge of C

The course topics included:

- Network architectures
- Protocols
- Media
- Software
- Measurement
- Security
- Politics

3.4.2 Example #2: Network Programming Course A more common course is one that meets software programming requirements but does not contain the hardware components. One example of a network programming course is taught by David Kuncicky at Florida State University [11].

The required prerequisites are:

- Working knowledge of the C programming language (at least two C or C++ programming courses)
- User-level knowledge of UNIX

Combinations of the following texts were used in various instantiations of the network programming course. No single text adequately covers all of the topics. Text material is supplemented with Web-based lecture notes.

- Stevens, W. R. *UNIX Network Programming.* Englewood Cliffs, NJ: Prentice-Hall, 1990.
- Stevens, W. R. *Advanced Programming in the UNIX Environment.* Reading, MA: Addison-Wesley Professional Computing Series, 1992.
- Quinn, B. and Shute, D. *Windows Sockets Network Programming.* Reading, MA: Addison-Wesley Advanced Windows Series, 1996.
- Comer, Douglas E. and Stevens, D. L. *Internetworking with TCP/IP, Volume III Client-Server Programming and Applications* (Windows Sockets Version). Englewood Cliffs, NJ: Prentice-Hall, 1997.

Other recommended reference texts are:

- Stevens, W. R. *TCP/IP Illustrated Volumes 1-3.* Reading, MA: Addison-Wesley Professional Computing Series, 1994.
- Tanenbaum, Andrew S. *Computer Networks, Third Edition.* Englewood Cliffs, NJ: Prentice Hall, 1996.

Typical programming assignments for the network programming course have been applications such as:

- An FTP server and client
- A chat server and client
- Network administration tools using SNMP

The course was first taught five years ago and was initially UNIX-based. Over time, the course content has expanded to include discussion of other platforms such as Microsoft Windows.

3.5 The System Administration Seminar

The system administration seminar is intended to provide a forum for presenting some of the components of system administration other than technical task-oriented skills. Suggested topics include:

- Ethics
- The role of the system administrator
- System administration related research
- Sharing project and thesis ideas (graduate level)
- Organizational models and issues
- Job trends
- Guest lectures by system administration professionals

A possibility is to offer this as a one-credit course in the final semester before graduation. This concept is being pursued informally at Florida State University in the computer and network system administration track. It has not yet officially been incorporated into the program.

3.6 The Experiential Component

There is consensus that the education of new system administrators requires an experiential component. Many fields within higher education require experiential training. In some areas such as engineering and chemistry, these experiences may take the form of laboratory courses. In other areas such as social work, education, nursing, and medicine, the required experience takes the form of an internship.

Almost all of the instructors who were contacted during the writing of this booklet mentioned the importance of hands-on experience as a component of learning system administration. Several of the instructors also mentioned that it was difficult to provide bona fide hands-on experiences in a classroom or even a laboratory setting.

The installation of a Linux box for a single user is not the same as a complex installation or upgrade on a live system. A real-world installation or upgrade might include a server that is integrated into a cluster of servers on a live network with hundreds of host computers and thousands of users. The installation of a software package on a single workstation is not the same as performing the installation on a large system where users may have developed dependencies on previous versions of software. The management of accounts in a laboratory setting doesn't include walking the political tight rope that system administrators learn to tread.

It is clear that a system administration program should have some form of extensive practicum. The practicum should include guided experiences that provide the student with exposure to real-world complexity and stresses. Possible ways to implement a practicum are through internships, the use of students to manage the department's labs, and the use of competency checklists.

3.6.1 Internship A traditional method for training professionals is a post-classroom internship or co-op program. This is common in the fields of social work, education, medicine, and nursing. An internship provides a guided, real world training experience that allows a novice to enjoy the tutorship of more experienced professionals. Employers view internships as a relatively safe way of screening potential employees. There are several reasons, however, why internships may not be as effective for the system administration profession as in helping professions.

- No licensure. Other professions hold the promise of licensure for students after graduation. For example, in some states social workers must be employed in a supervised setting for three years after graduation before being certified. There is already a problem with system administration students leaving programs for employment before graduation. A student would certainly be tempted to forego an internship if a high-paying job is available that doesn't even require a degree.

- Cost. An internship program requires a great deal of human resources to administer correctly. Office staff is needed to administer the program. Relationships with the recipient institutions must be cultivated. Many internship programs require faculty members to spend time on-site to supervise and assess students. The department that is already reluctant to add system administration to its curriculum is not likely to be willing to pour resources into an internship program.
- Liability. An official, sponsored internship program means that the sponsoring department may have an implied or explicit liability to a company or institution that allows the student and/or instructor to have root level access to their computing resources.

3.6.2 Competency Checklist One idea that has been discussed recently within the system administration community is the use of competency checklists, or what has been dubbed the "merit badge" approach. The intention is to use competency checklists as a means of validating and maintaining consistency of real-world experiences. These could be used as an alternate and/or supplement to an organized internship program.

For those of you who have not been exposed to Boy Scouts here is a brief description of merit badges recalled from 37 years ago. One of the ways to advance in rank within scouts is to collect a number of merit badges that are organized in various categories. Some examples of merit badge areas are cooking, camping, archery, and hiking. Each merit badge has an associated manual that describes the subject area and states the requirements that must be completed in order to obtain the badge. The requirements can consist of written exams and written reports but frequently consist of performance requirements, e.g., cooking bread over an open campfire. These are learned under the tutelage of an approved mentor who then signs off when the tasks have been successfully completed. The process of earning the merit badge is essentially a mini-apprenticeship for a single narrow domain. The responsibility for seeking out an approved tutor belongs to the student. Any apprenticeship agreement is solely between the tutor and the student, which helps to resolve the cost and liability issues of internships. The issue of competency checklist development is explored further in the following section.

3.7 Competency Checklist Development

The notion of the competency checklist is in its infancy. This is an area where SAGE could take a leadership role. One suggested way to approach the development of a system administration competency checklist is to create a taxonomy of system administration skills. The highest abstract levels of skills are called competency domains. Following are examples of some suggested competency domains:

- Hardware maintenance
- Booting
- System backup

- File systems
- System programming
- Mail services
- Networking
- Printing
- Software installation
- Peripheral management
- Name service
- User management
- Security
- Performance analysis
- Administrative duties

Competency domains are relatively unchanging categories that may contain a single competency (e.g., booting) or may represent a collection of competencies. For example, the file systems domain might contain the following competencies:

- Installation of a file system
- Disk installation
- Performance tuning
- File sharing
- Disk space management

For each competency, a collection of skills is stipulated. This collection of skills must be learned under an approved apprenticeship. Examples of skills to be learned and performed for the file sharing competency are:

- Set up NFS service on a UNIX file server
- Set up an automounter on a UNIX file server
- Set up Windows/95 and Windows NT volume sharing
- Set up folder sharing on a Macintosh
- Install samba on a UNIX server for UNIX/Windows file and printer sharing
- Install netatalk or CAP on a UNIX server for UNIX/Macintosh file and printer sharing

The specific skills within a competency will tend to change more rapidly than the domain itself. Specific requirements will have to be regularly updated. One method that would give the requirements a longer lifetime is to offer choices of requirements to candidates. For example, one could require that four of these six tasks must be completed to satisfy the file sharing competency. The list would still be usable if Apple went out of business.

A competency-based approach has some advantages over an internship approach. A competency-based approach is student-driven. That is, the student must seek out an approved tutor to satisfy the requirements of the competency. The tutor is more likely to be a real expert for the specific domain in question. For example, a student might take an internship with a company or department for six months and never get the opportunity to configure a router. The campus network coordinator, on the other hand, could administer the network management competency. The student would likely receive competent training and would have the opportunity to learn router configuration from a specialist. The expert's resources are more efficiently used under this method. The costs to the degree-granting department should be considerably lower than the costs of maintaining an internship program.

Care should be taken to require as much of a real-world environment as possible. To prohibit students from satisfying all competencies on a single-user Linux host, additional qualifications can be made for certain competencies. Examples of qualifications are:

- The server must have some specified minimum number of active accounts (e.g. 50 accounts).
- This competency must be performed on an active server that requires scheduled downtime.
- This competency must be performed on at least four operating systems (e.g., Linux, HP-UX, Solaris, NT, BSD).

The competency checklist method does not preclude part-time employment or an internship during the program. In fact, employment may be offered in graduate programs in the form of assistantships. This method is used at FSU (see the case study). The student is provided a small salary and tuition deferment to work in a campus department or research institute that has a sufficiently rich set of computer and network devices. With little overhead, the department can facilitate matching students with jobs in which they can gain appropriate experience.

4.0 Commercial Training

4.1 Introduction

This section provides a survey of system administration training and educational opportunities outside academia. Such training has a long history. Many major corporations have their own well-developed education departments. For example, the training center developed by Andersen Consulting in St. Charles, Illinois, houses a campus that would be the envy of many colleges. The site is on a 151-acre campus that includes classroom buildings, residence halls, dining facilities, and top-flight athletic facilities including 35 miles of bike and fitness trails!

In spite of such facilities at major corporations, many companies are not able to keep up with the training needs of their employees and clients. Many of these companies outsource some or all of their educational needs to third-party vendors who specialize in training and development. More recently, universities and corporations have been seeking each other out and exploring ways to form partnerships to develop and deliver educational courseware in the areas of computer and network system administration. The following sections discuss each of these forms of commercial training and provide examples of each.

4.2 In-House Training

An education department within a corporation has two major functions: orientation/training of new employees and retraining for new technologies, i.e., continuing education. Providing training in house is expensive. The costs include educational staff, classroom space, training labs, and the expense of covering employees' time during their absence. The cost/benefit ratio for this form of delivery is generally lowest when the content is site-specific or if the technology is developed or manufactured at the company in question.

4.2.1 Orientation Methods of delivery for in-house training include classroom settings, video-conferencing, Web-based tutorials, and on-the-job training. Classroom lectures and written exercises can be augmented by practical experience in a laboratory setting or by on-the-job training. If company experts are involved with the delivery of the training, then students begin to meet future mentors during the orientation process. If the corporation has an internship program, then the resources used to train interns can be combined with the other resources of the education center.

An orientation program may begin with a series of intense classroom training courses. The employee then typically works alongside a designated mentor in the workplace in order to practice applying the newly learned knowledge. Finally, the orientation program may include an evaluation process to verify that the new employee has successfully learned the skills required for the system administrator role.

In some companies, the new employee may rotate through a number of work assignments and mentors, depending on the size of the organization. If there is a well-defined set of skills to be mastered, then mentors could use a type of competency checklists to track success. Costs can be minimized by preventing duplication of training from one assignment to the next. The employee determines the length of the training by mastering each competency at the employee's own pace.

4.2.2 Continuing Education It is critically important that a corporation provides a means for its technical staff to learn emerging technology. The frequent need for retraining is exacerbated by high employee turnover in the information technology field. Retention of staff is a major problem for high technology companies today. At this time, the average length of employment at a single site for a system administrator is 2.5 years. The promise and delivery of continuing education is one method of improving morale, providing growth opportunities for technical staff, and possibly increasing employee retention. An in-house education department generally cannot meet the complete continuing education needs of system administrators. The corporation should also utilize external resources such as vendor-based training, third-party trainers, local college and universities, and technical conferences.

4.3 Vendor-Based External Training

Many vendors of computing and networking systems provide training for their clients in addition to their own employees. This is called vendor-based external training. The incentive for corporations is obvious. Customers demand that there be enough qualified system administrators in the market to manage the corporations' products. For some corporations, the training service is a profit center. The vendors' customers benefit from a reduced overall cost of ownership and accelerated deployment capability. Vendor-based training is narrowly focused by design. It is more useful if the student already has general skills and knowledge in the presented topic.

There are too many examples of this type of training to provide a comprehensive list of vendors. Here is one example from The Open Group's Distributed Computing Environment (DCE).

The Open Group is an international consortium of vendors and end users in industry, government, and academia. The Open Group includes the Open Software Foundation and X/Open Company, Ltd.

In 1996, the Open Group began the design and implementation of a DCE skills certification program targeting DCE system administrators. The skills certification program concentrates on the skills and knowledge required for the administration of DCE core services. According to Peter Shaw, OSF's vice president of sales and marketing, "OSF is firmly committed to ensuring that the skills necessary to make DCE successful in the marketplace are available through this key technology transfer program. We see DCE skills certification as a logical extension of our platform certification program and we look forward to broad participation in this program from both end user companies and the computer industry [12]."

The Open Group offers the DCE Administration Skills Certification tests through a worldwide network of certification testing centers. John Raleigh, OSF's Worldwide Professional Services Program manager, acknowledges that a critical shortage of qualified DCE professionals exists, and that certification programs appear to be the best way for DCE to satisfy its mission-critical IT requirements [12].

4.4 Third-Party Trainers

A number of companies specialize in the training and education of computer and network system administrators. These companies tend to provide vendor-specific training (e.g., HP-UX, Novell, and Microsoft NT). The better providers include laboratory experience as part of the training and, upon graduation, the student can expect to be prepared for a vendor's certification exam.

The courses are typically taught within a short time frame (three to five days) and can be expensive, especially if travel and lodging costs are required. Nevertheless, this type of course may offer a corporation the best means of providing its employees training in a new technology, especially if only a few employees will need the training. If many employees need to be trained, some third-party trainers will provide on-site courses. An alternative is to send employees away to be trained as trainers.

4.5 Corporate-Academic Partnerships

As the gap between supply and demand of competent and well-trained system administrators grows, some corporations are proposing strengthened partnerships with higher education. For such a partnership to work, both parties must be willing to bridge the long-term differences in their approaches to training. The corporation cannot expect to control course content, and the college or university cannot expect to teach materials irrelevant to the corporation's needs. One method of achieving a balance is to have a higher-education institution offer a course or program of courses that prepares the student for a certification exam. The exam is administered and controlled by the corporation. The professor designs and controls the course. The course content may have broad scope and the course does not teach to the exam. Preparation for the certification exam is a side effect of taking the course. The corporation can gauge the effectiveness of the course by the success rate on the certification exam.

Why would a school be interested in forming such a partnership? One reason is that the students become more marketable with certification. A second reason is that the corporation can assist a department or college with the development of teaching laboratories. During this period of tightened funding at many state universities, departments are having a difficult time building and maintaining current, high-tech teaching labs for hands-on training. A third reason that corporate partnership helps schools is that corporate sponsorship can offer exposure for new emerging school programs. Finally, internship programs can be implemented that benefit both the higher-education institution and the corporation.

One example of a partnership of this nature is between Cisco Systems and Dalhousie University in Halifax. Beginning in the fall of 1997, DalTech will offer a new, one-year interdisciplinary Master of Engineering in Internetworking. One of the sponsors is Cisco Systems [13].

Cisco is experiencing a worldwide shortage of skilled network administrators. The highest level certification that Cisco offers is called the Cisco Certified Internetwork Engineer (CCIE). Cisco designed the CCIE program as a high-level certification vehicle to ensure that its customers as well as its own technical staffs have the expert level of internetwork knowledge required to manage its installed base of routers and switches. About 3,000 individuals are CCIE-certified worldwide. Applicants sitting for the CCIE exam suffer a 60% failure rate. The exam includes a grueling hands-on laboratory component.

The Master of Engineering at DalTech has been designed to prepare individuals to play an active role in the rapidly expanding field of Internetworking. The program is open to students with an undergraduate degree in electrical engineering, computer engineering, industrial engineering, or computer science. The program accepts non-degree-seeking students as students in individual courses. This will benefit local professionals who wish to upgrade their knowledge and skills in a specific area.

The leaders emphasize that they offer more than a diploma or certification program. The Master's degree curriculum covers not only how to use current technology, but also why the technology is in its present form. "The program provides the theoretical background to analyze the shortcomings and strengths of the technology, its continuing evolution, and the challenges that lie ahead for the industry." [13] The degree requirements consist of ten courses and a project scheduled over a 12-month period. One distinction of this program when compared to most other graduate programs is that the courses will be offered in a compressed two-week format that can be taken on either a full-time or a part-time basis. Practical experience is to be acquired through internships that are arranged with sponsoring companies.

4.6 Professional Conferences

Attendance at technical conferences is another method by which employees may receive training. In addition, conferences provide the system administrator with the opportunity to hear about upcoming technologies and share experiences with colleagues.

Conferences often provide tutorial sessions in which attendees can learn from an expert in a given area. These presentations are useful for obtaining a general introduction to a subject area that is new to the attendee. Conference tutorials generally do not restrict or pre-test attendees. They generally do not incorporate a hands-on component, and they do not usually provide any form of assessment or certification. These drawbacks are offset by the opportunity for the student to learn from real experts in the field and, in many cases, the original authors or developers of the topics being taught. For example, if one wants to learn about sendmail issues, who better to learn from than Eric Allman, the author and maintainer of sendmail. The primary technical conferences for system administrators include LISA [14], USENIX [15], and SANS [16].

5.0 Independent Study

5.1 Introduction

System administrators tend to be independent learners out of necessity. Pressure is increasing on corporations and institutions of higher education to explore independent study programs. These programs are grouped under names such as distance learning or video-based training. Some might argue that distance learning is not synonymous with independent study since teacher-student interaction does occur. Our experience is that the successful completion of distance learning courses requires a higher level of self-motivation than does classroom-based education. In any case, distance learning courses are independent in the sense that the student may take them independent of place or time.

Corporations are finding themselves unable to keep up with the training demands of their clients. As mentioned in the previous section, companies resolve this problem by outsourcing to third-party trainers and collaborating with institutions of higher education.

Institutions of higher education are also experiencing growing pains. Enrollment is increasing rapidly as the children of the baby boomers reach college age. However, the construction of new buildings and the development of other infrastructure are not keeping pace. One method of counteracting this trend is developing independent study and/or distance learning programs.

The most rapidly growing delivery method for independent study programs is the use of Internet-based technology. WWW-based, hypertext delivery mechanisms have several advantages.

One is pedagogical. There is some evidence to support the notion that user-driven hypertext documents are highly appropriate for learning in complex and ill-structured domains [17]. Computer and network system administration is definitely a complex and ill-structured domain. It is a difficult subject to teach in a straightforward linear fashion. Hypertext allows a nonlinear and multidimensional traversal of complex subject matter. The Web-like structure of hypertext course material can more accurately reflect the Web-like structure of the real knowledge domain. Case studies may be developed that allow multiple paths to be taken through the cases. The threads can be used to demonstrate different themes throughout the course. The access and flow of the instruction is learner-controlled. This allows the learner to pause, reflect, and assimilate knowledge at the learner's own pace.

The ability to modify and update the subject matter is a pragmatic advantage to Web-based courseware. Course revisions are immediately visible to learners. Web-based courseware may be less expensive than other forms of delivery, but that claim has yet to be settled definitively.

Disadvantages to all methods of independent study are the difficulties in administering evaluations and delivering practical training. Some programs offer supplemental lab training and/or on-site evaluation conjunction with Web-based course delivery. The USAIL project from Indiana University uses this method, and it is discussed later. Independent-study degree programs exist that resolve these issues. The Regent's College External Degree Program from the State University of New York will also be discussed. Finally, one of the case studies presented describes Pencom's efforts at Web-based delivery methods.

5.2 UNIX System Administration Independent Learning (USAIL)

One example of a recently developed independent study program in system administration is the USAIL project at Indiana University. The USAIL project is part of a larger effort by the University Computing Services at Indiana University to provide computing education programs that lead to certification. Interested readers may want to visit the USAIL Web site [18], the EdCert Web site, or download a paper by Raven Tompkins that discusses the USAIL project and was presented at LISA X [19].

The USAIL and EdCert programs address many of the difficult problems that face educators who are devising independent study programs in system administration.

5.2.1 Pre-test The program lists the skills that are required as prerequisites to the course. A pre-test is administered that gives potential students feedback on the appropriateness of the course for their knowledge and skill level. Based on the results of the pre-test, highly qualified students can take the EdCert final exam without receiving any further instruction.

5.2.2 Web-based course delivery The usual benefits of Web-based delivery apply. Students have asynchronous access to Web-based materials. Students may move through the coursework and take the quizzes at their own pace. From the educator's point of view, an advantage is that materials may be easily updated and revised.

5.2.3 On-line evaluation Students may take all written quizzes except the final exam on-line and at their own pace.

5.2.4 Lab-based practical experience There are nine on-site lab-based classes for the course. During these labs, the students get hands-on tutelage from instructors.

5.2.5 Written and practical lab-based evaluation The final exam is a proctored exam that consists of both written questions and hands-on work.

The concept could be expanded into a series of courses. A nonacademic department administers the program. Thus, academic credit is not awarded at this time. In principle, the concept could be used as the basis for a degree-granting program.

5.3 Regent's External Degree Program.

During the 1970s, a few educational leaders became convinced that it should not matter where or how students acquire their knowledge. If students' knowledge and skills are the same as that of college graduates, the student should be entitled to the same recognition, the college degree. These educators began creating a college system for students who are willing to acquire knowledge and skills independently [20].

One of the most successful of these programs is the Regent's College External Degree Program at the State University of New York. The Regent's nursing program serves as an illustration of a potential external degree program in system administration. The education of professional nurses has many parallels to the education of system administrators. Professional nursing requires a high level of technical competence, the ability to multitask under pressure, effective communication skills, problem-solving skills, and political shrewdness. The technical knowledge domain changes rapidly. Nursing students must have some sort of practicum during their training to hone their skills and gain experience before they graduate. There are obvious differences. Nurses make decisions that have life-or-death consequences. The technical domains are obviously different. Nevertheless, the types of demands and general requirements are very similar to those of system administrators.

The Regent's program has been graduating nurses with external degrees for more than 20 years. The graduates may sit for nursing boards and be licensed as Registered Nurse in most states.

Students are required to obtain study materials, learn the materials, and then pass written exams for the knowledge-based component. For the skill-based component, students are required to set up their own informal internship, or in some other manner, students must learn the procedural skills required of a professional nurse. The final exam consists of several days of real nursing assignments in a hospital.

An external degree program in system administration could be based on this model. The development of a thorough competency checklist could serve as a way of defining and standardizing the required set of procedural skills. SAGE is a natural organization to provide leadership in this area.

6.0 Case Studies

6.1 Introduction

Two case studies are presented in the following sections, one from academia and one from the commercial sector. The case study from academia is the masters program in computer and network system administration at Florida State University. The case study from industry is Pencom's internal Web-based training program. There are at least two other academic programs of note: one at Central Queensland University and the other at the University of Colorado at Boulder. Each of these programs represents an attempt to build system administration into the curriculum of the respective department in a substantial manner. The programs each require the completion of experiential components for graduation.

6.2 Case Study 1: FSU Master's Program in Computer and Network System Administration

The Department of Computer Science at Florida State University administers a Computer and Network System Administration Masters Track [21].

6.2.1 History As is the case in many departments of computer science and computer engineering, the computing and network system is managed by a small group of students. The work may be done under the guidance of a faculty member, but the students perform much of the day-to-day work. This situation existed in the late 1980s at Florida State University. At any one time, three or four graduate students would run the system as a way of paying off their assistantship. The students would simultaneously write a thesis in some area of computer science other than system administration. Inevitably, each student would graduate and would take a job not in their field of research but as a system administrator. The faculty members decided that they should simply formalize the process that they were observing and designed a master's track in computer and network system administration.

Several new courses were added to the graduate curriculum. They were taught first as special topic electives, and then as core courses in a newly declared track. The program was small at the time. Between the computer science department and the nearby FSU Supercomputer Computations Research Institute, the program could ensure an experiential component for students by hiring each of them as a student system administrator. After the program accepted more than about eight students, this method was no longer viable.

Since the program's inception, the number of faculty members willing to accept majors in this track has grown from two to nine. Seven students have completed the program and 12 more have been accepted into the track. The faculty purposely restricted access to the program in the first two years to tune the requirements and test the curriculum. The program now can maintain 15-20 full-time students.

An early problem mentioned by the faculty was difficulty in finding acceptable project or thesis topics in this area. A great deal of internal discussion ensued and it was decided that an oversight committee would review thesis and project topics. This process helped to insure some consistency among projects and fostered communication among the faculty members and among the students. Attendance at USENIX and SAGE conferences and review of SAGE and USENIX conference proceedings provided a wealth of information for students who were looking for appropriate topic content and project scope. In addition, the CNSA students regularly gather to discuss their projects and other system administration topics. These meetings help give students a chance to present their ideas and obtain feedback in a nonthreatening environment.

A second problem has been the development of a consistent method of providing guided practical experience for students. As the program grew, it became difficult to hire all of the students. The method devised to replace hiring each student was the competency checklist, described in detail in an earlier section. The competency checklist is a work in progress. One continuing problem with the competency checklist method is the recruitment and retention of competent experts. Another is the continued definition and revision of competency domains and the specific required skills within those domains.

6.2.2 Graduation Requirements The Computer and Network System Administration major has the same admission requirements as the general Computer Science major, together with a pre/co-requisite experience base that includes work in a system administration environment. Admission to the Computer and Network System Administration major is selective and limited, and is managed by the System Administration Faculty Oversight Committee.

In the Computer and Network System Administration major, the student must complete at least 32 semester hours in computer science courses numbered 5000 or above, including the courses:

- CDA 5155 Computer Architecture (3 credit hours)
- CEN 5515 Data and Computer Communications (3 credit hours)
- CIS 5406 Computer and Network Administration (3 credit hours)
- CIS 5935 Introductory Seminar on Research (1 credit hour)

- COP 5570 Advanced Network Programming (3 credit hours)
- COP 5611 Operating Systems (3 credit hours)

6.2.3 The System Administration Competency Checklist The Computer Network and System Administration track requires that each student have practical experience performing tasks that demonstrate the competency expected of a professional in this area. This requirement cannot be satisfied by coursework alone, but must involve work as administrator of a networked computer system with real users. It is a condition of admission to the track that each student agrees to take responsibility for obtaining the required experience, and demonstrate that the requirement has been satisfied, before graduation. In some cases, a student may enter the program with qualifying prior experience. In other cases, students must seek opportunities to gain system administration experience between the time of admission and graduation. These may be paid or voluntary positions. For graduation, the student is required to demonstrate at least one of the specified skills in each competency on the competency checklist. As the competency checklist undergoes changes, each revision is dated and archived. This practice allows a student to follow a single consistent program and, at the same time, allows the department to revise the checklist and maintain a current skill set.

6.3 Case Study 2: Pencom's Internal Web-Based Training

Early in 1995, Pencom's system administration team recognized a need for a flexible training program that would allow knowledge gained in the field to be shared among the growing workforce of UNIX systems administrators.

In order to organize the skills and education into a meaningful curriculum, it was first necessary to determine some type of metric that could be used to measure a systems administrator's level of expertise. It was also necessary to identify the types of technologies that are required to provide systems and network administration support in the field.

At about this time, SAGE published *Job Descriptions for Systems Administrators.* This booklet became the foundation for Pencom's training curriculum because it provided a logical progression of the level of work performed by increasingly experienced UNIX system administrators.

Using *Job Descriptions* as a guideline, members of Pencom's technical staff created similar skill set descriptions for system engineers of Microsoft NT and expanded the UNIX skill sets to include more extensive network administration skills. A special working group, called Pencom Systems Administration University, or PSAU, was formed.

The PSAU defined a training curriculum including courses that would be required to achieve the PSAU Junior through PSAU Senior levels of UNIX and NT administration. This involved expanding the skill sets described in *Job Descriptions.* Course descriptions and measurable performance objectives were written. To provide career planning, the skill levels were incorporated into PSA's employee-development planning.

When the time came to begin looking at the proposed courses themselves, it was discovered that much work was needed to plan an appropriate delivery mechanism. Since Pencom's workforce is distributed over the United States and other countries, the traditional classroom presentation model would not work. Pencom employees have such varied skill requirements that there were usually only a small number of employees seeking training at a particular site in a single topic at any one time.

It was decided to develop the curriculum as a series of Web-based tutorials. A template for interactive courseware was built that delivers lectures, provides exercises, and records students' progress. Password protection restricted the review of a student's records to the student and the respective manager.

Over time, it was discovered that students benefited more from exercises that required them to leave the Web page and work on an actual machine in a real environment. Many exercises were modified to include lists of specific commands to use on a real machine.

One drawback to Pencom's Web-based approach is that it is limited to the command line interface. Training modules for systems that use GUI system administration tools are more challenging to design and slower to download and use.

A major benefit of a Web-based approach is the opportunity to learn asynchronously. The student may learn independent of place and time. On-line coursework allows electronic record keeping, which is beneficial for students and managers.

7.0 Goals and Directions

This booklet has presented some of the challenges faced by students, educators, and employers in educating and training competent system administrators. It has also described the state of educational programs in this area. It represents the beginning of a discussion that needs to be continued. To meet the present and expected demand for system administrators in the next decade, a number of goals must be met. These include:

- The implementation of new delivery paradigms for educating system administrators
- Wider collaboration between educators and industry
- Wider acceptance of system administration track studies at the undergraduate and graduate level

SAGE could act as a catalyst and facilitator among the principal players: computing vendors, third-party trainers, institutions of higher education, and other professional organizations. Several examples will illustrate potential roles for SAGE.

One example lies in the academic realm. The four or five programs offering advanced degrees in computer and network system administration have had little inter-institution communication. Some of the faculty members have expressed surprise that the other programs even exist. There is a need for educators to share experiences and to work together on topics such as curriculum development. The sponsorship of workshops for system administration educators may be appropriate as the formalization of programs in this area continues. An outgrowth of the workshops might be a prototype SAGE-approved curriculum for several levels of degree programs in system administration. These could be comprised of detailed guidelines for curriculum development such as course descriptions, lecture outlines, exercises, suggestions for the development of internship programs, and descriptions of competencies.

Another area in which SAGE could play a role is certification. There are many examples of vendor-specific certification, but there has been little development of certificate programs that span platforms and cover domain-based topics. If certification were to come about as part of this profession's evolution, SAGE could certainly be an unbiased focal point and leader in the development of competency-based certification for system administrators.

Appendix: Web-Accessible Courses, Programs, and Resources

The following links point to Web pages for some of the courses currently offered in the area of computer and network system administration. Links to the graduate and certificate programs mentioned in the booklet are also listed. URLs tend to move over time. These links were accessed successfully on September 10, 1998. A current version of these and other related resources may be found at *http://www.eng.fsu.edu/saed.* Please send additions and corrections to Dave Kuncicky at *kuncicky@eng.fsu.edu.*

Selected Courses from Academic Institutions

Computer and Network System Administration (CIS 5406), Summer 1998
http://www.cs.fsu.edu/~jtbauer/cis5406/index.html
Florida State University
Jeff Bauer, *<jtbauer@cs.fsu.edu>*

UNIX System Administration (OSU CS312), Summer 1997 *http://iq.orst.edu/sysadm/*
Oregon State University
John Sechrest, *<sechrest@cs.orst.edu>*

UNIX System Administration - A Survival Course
http://www.washington.edu/R870/
University of Washington
Dave Dittrich, *<dittrich@cac.washington.edu>*

Introduction to UNIX Systems Administration (IFSM 498B)
http://www.gl.umbc.edu/~jack/ifsm498d/
University of Maryland, Baltimore County
Jack Suess, *<jack@umbc.edu>*

Certified Sun Training, 1998
http://www.hcc.hawaii.edu/matsuda/sun.html
Honolulu Community College
Roy Inouye, *<sun-contact@hcc.hawaii.edu>*

Systems Administration (85321), Fall 1998
http://infocom.cqu.edu.au/85321/
Central Queensland University
David Jones, <*d.jones@cqu.edu.au*>

UNIX Systems Administration (CS 49995/59995), Spring 1997
http://dune.mcs.kent.edu/~farrell/sa96/index.html
Kent State University
Paul A. Farrell, <*farrell@mcs.kent.edu*>

Educational Programs in System Administration

Computer and Network System Administration Masters Track in Computer Science
http://www.cs.fsu.edu/system_administration/cnsa-index.html
Department of Computer Science, Florida State University

Master of Engineering in Internetworking at DalTech
http://is.dal.ca/~eine/
College of Applied Science and Technology within Dalhousie University, Halifax, Nova Scotia

USAIL: UNIX System Administration Independent Learning
http://www.uwsg.indiana.edu/usail/
Indiana University

UNIX System Administration Certificate Program
http://www.cc.gatech.edu/conted/unix.certificate.html
Georgia Tech, College of Computing

Selected Vendor-Based Training Sites

Hewlett Packard Network and System Administration Courses
http://www.hp.com/pso/frames/services/ed-services.html
Hewlett Packard

Sun Educational Services
http://www.sun.com/sunservice/suned/
Sun Microsystems

Configuration of NCD Network Computers (UNIX)
http://www.ncd.com/support/services/training.html
Network Computing Devices

Sequent UNIX Curriculum Course Descriptions
http://www.sequent.com/offerings/services/educ_svcs/text/catalog/unix.html
Sequent Corporation

Selected Third-Party Training Sites

miSoft Systems Engineering & Solaris Training
http://www.misoft.com/
miSOFT

Unix Network and System Administration Training
http://www.compclass.com/
The Computer Classroom, Inc.

Multi-Vendor System Administration Training
http://www.metroinfo.com/train/
PPR Corporation

Network and System Administration Training
http://www.xor.com/
XOR Network Engineering

Other Resources

Training Courses for System Administration
http://www.stokely.com/unix.sysadm.resources/courses.html

References

1. Darmohray, T., ed. *Job Descriptions for System Administrators*, 2d ed. Berkeley, CA: USENIX Association, 1997. <*http://www.usenix.org/sage/jobs/jobs-descriptions.html*>.
2. Nemeth, E., G. Snyder, S. Seebass, and T. Hein. *UNIX System Administration Handbook*, 2d ed. Englewood Cliffs, NJ: Prentice Hall, 1995.
3. Jones, D. "How do you teach System Administration?" *Proceedings of the Seventh Systems Administration Conference (LISA VII)*, Monterey, CA, 1993. <*http://cq-pan.cqu.edu.au/david-jones/Publications/Papers_and_Books/93sage/*>.
4. Jones, D. "Teaching Systems Administration II" *Proceedings of SAGE-AU'95*, Wollongong, 1995. <*http://cq-pan.cqu.edu.au/david-jones/Publications/Papers_and_Books/95admin2/*>.
5. Jones, D. *85321, Systems Administration* [Web page], Central Queensland University, Summer 1997. <*http://science.cqu.edu.au/mc/Academic_Programs/Units/85321_Systems_Administration/*>.
6. Frisch, A. *Essential System Administration,* 2d ed. Sebastopol, CA.: O'Reilly and Associates, 1995.
7. Sobel, M. *A Practical Guide to the UNIX System*, Redwood City, CA: The Benjamin/Cummings Publishing Company, Inc., 1995.
8. Glass, G. *UNIX for Programmers and Users, A Complete Guide*, Englewood Cliffs, NJ: Prentice Hall, 1993.
9. Bauer, J. *CIS 5406 Computer and Network System Administration* [Web page], Florida State University, Summer 1997. <*http://www.cs.fsu.edu/~jtbauer/cis5406/index.html*>.
10. Nemeth, E. *CS 4273/5273 Network Systems Course Description* [Web page]. <*http://www.cs.colorado.edu/homes/evi/public_html/networks/announce.html*>.
11. Kuncicky, D. *COP 5570/ ENG 5930 Advanced Network Programming* [Web page], Florida State University, Fall 1996. <*http://www.eng.fsu.edu/~kuncick/unixcourse96/*>, Spring 1998, <*http://www.eng.fsu.edu/net98*>.
12. *Distributed Computing Environment (DCE)* [Web page]. <*http://www.opengroup.org/tech/dce/*>.

13. *Master of Engineering in Internetworking* [Web page], DalTech College of Applied Science and Technology, Dalhousie University, Halifax, Nova Scotia. <*http://is.dal.ca/~eine/*>.
14. *USENIX - LISA* [Web page]. <*http://www.usenix.org/events/bytopic/lisa.html*>.
15. *USENIX Annual Technical Conference* [Web page], <*http://www.usenix.org/events/bytopic/usenix.html*>.
16. *SANS* [Web page], <*http://www.sans.org/*>.
17. Spiro, R. J. and J. Jheng "Cognitive Flexibility, Constructivism, and Hypertext: Random Access Instruction for Advanced Knowledge Acquisition in Ill-Structured Domains," in T. Duffy & D. Jonassen, (eds.), *Constructivism and the Technology of Instruction*, (pp. 57-75). Hillsdale, NJ: Lawrence Erlbaum Associates, 1991. <*http://www.ilt.columbia.edu/ilt/papers/Spiro.html*>.
18. *USAIL: UNIX System Administration Independent Learning* [Web page], Indiana University. <*http://www.uwsg.indiana.edu/usail/*>.
19. Tompkins, R. "A New Twist on Teaching System Administration," *Proceedings of the Tenth Systems Administration Conference (LISA X)*, Chicago, IL, 1996. <*http://www.uwsg.indiana.edu/usail/lisa.html*>.
20. *Regent's College* [Web page], The State University of New York. <*http://www.regents.edu*>.
21. *Computer and Network System Administration Masters Track* [Web page], Department of Computer Science, Florida State University. <*http://www.cs.fsu.edu/academics/sysadm/*>.